The Evolution of Condemnation

Jonathan Leigh, M.D.

Sage Book Publishing

Los Angeles

Published by Sage Book Publishing, Los Angeles

1 2 3 4 5 6 7 8 9 10

Prologue

Evolutionary psychology is not a branch of psychology: it is the base of psychology. All psychology must take into account, either explicitly or implicitly, the evolutionary origin of the psychological mechanisms that produce human behavior. As a discipline, evolutionary psychology focuses on explaining human behavior in terms of these evolved psychological mechanisms, as well as elucidating their evolutionary origin. If these psychological mechanisms evolved, then they are genetically-encoded biological adaptations. In evolutionary biology, an adaptation is an evolved trait shaped by natural selection that promoted an individual's survival and reproductive success in the evolutionary past of the species. Evolutionary psychology has successfully explained many human behaviors as the product of biological adaptations, including love, jealousy, aggression, cooperation, and friendship. Even though psychology is recognized as a branch of the social sciences, it is in reality an extension of biology, and therefore should be classified as a natural science, similar to chemistry, physics and cosmology. This should have been made clear by the success of the neurosciences in explaining human behavior, in addition to the success of evolutionary psychology. Indeed, other social sciences, such as history and anthropology, should be recognized and accepted as extensions of biology, and therefore considered natural sciences as well. In other words, there should no longer be any sharp distinction between the social sciences and the natural sciences. (Naturally, the Humanities would be the next disciplines to be encircled within the folds of Science.)

In this book, evolutionary psychology is employed to elucídate a human phenomenon that has so far been highly resistant to scientific inquiry: morality. Piece by piece, the edifice of human morality is explored and scrutinized under the rigors of science. At the end of the book, human morality will have been explained in purely scientific terms.

Chapter 1: Condemnation Throughout Human History and Across Cultures

To explain morality, one needs to examine an aspect of human behavior that pervades all moral systems—condemnation. It will be argued facilely that human morality is more than just condemnation, but if one delves further, one finds that condemnation is actually the foundation of all moral systems, without exception. If a moral dictum, precept, or code exists in any group or culture, it is implied that condemnation will be expected and deserved if there is a violation of it, with no explanation or justification needed. This may raise the question of why anyone would violate a moral dictum in the first place. From an evolutionary psychologist's perspective, the answer is obvious and straightforward: The violation of the moral dictum is compelled by an evolved psychological mechanism and is perceived to be advantageous by the violator; the others do not approve, and therefore condemn the violator because the violation is perceived as disadvantageous to them. Some would argue that morality is not only about what is wrong, and therefore worthy of condemnation, but also about what is right, and therefore worthy of commendation. In fact, the point of deciding "what is right" is so that any deviation from it can be condemned.

What follows, then, is that to understand morality, we need to understand condemnation.

Let's define what condemnation is. Condemnation is the accusation that someone is harmful and therefore deserves punishment. It is a signal and a behavior, and it is a biological adaptation. Any biological adaptation evolved by natural selection is found in almost all members of the species. If the behavior of condemnation—or, to be more precise, the psychological mechanism behind the behavior of condemnation—is a biological adaptation, it must exist in all cultures and throughout history and prehistory, as long as there were members of the species *Homo sapiens*.

Although condemnation purports to inform an audience of the harmfulness of the condemned, the people who were the target of condemnation are often revealed to be harmless, making the condemnation false. This does not mean that the condemnation was issued mistakenly, but rather that condemnation has other functions besides informing people of the harmfulness of the condemned. As we will learn later, if condemnation is an evolved adaptation, then it should be beneficial to the condemner. Whether it benefits an audience, such as alerting them to the potential harm from the condemned, is secondary.

Examples of Condemnation

Let's examine the most famous cases of condemnation in history.

Socrates

Socrates was a seeker of truth and wisdom who liked to speak with anyone in Athens, including the prominent citizens who held significant power in the democratic state, to argue and debate various matters on topics such as ethics, politics, and physics. He was highly insistent on getting to the bottom of the problem under discussion, and this often led to the revelation that his interlocutors knew little or nothing about the topic at hand, despite their prior reputation of being respected experts. Since these discourses often took place in public with a gathering of curious listeners, many prominent citizens had been publicly shamed by Socrates when his questioning exposed their ignorance and pretense of knowledge and wisdom.

In 399 BC, Socrates was brought to trial before a jury of a few hundred male Athenians chosen by lottery. He was accused of two crimes: impiety and corruption of youth. His defense of himself, recounted by Plato in *Apology*, made it abundantly clear that he was guilty of neither. He was voted guilty by the Athenian jury anyway. Some of them might have been insulted by Socrates in one of his many truth-seeking interrogative exchanges or were friends of other citizens thus humiliated, and the guilty vote served as a payback for them.

Others might have genuinely believed in the charges against Socrates. However, almost as many found him innocent as found him guilty; another 30 men voting in his favor would have acquitted him. This means that there was no consensus among the Athenians regarding Socrates' guilt.

After Socrates was convicted of his crimes, the prosecutor recommended death, but Socrates was allowed a chance to propose an alternative punishment. Indeed, many of those among the jury who voted for the guilty verdict might have believed that Socrates would choose excommunication from Athens as the alternative punishment, one that they would accept because it would eliminate his offensive presence from their lives forever. However, at the urging of his friends and students, Socrates offered money that his students would procure for him. The jury voted for the death penalty. Socrates drank hemlock bravely, leaving behind an inspiring portrait of the dauntless philosopher unwavering in the face of death.

Galileo

Galileo was an astronomer in Europe during the European Renaissance who observed other planets of the solar system revolving around the sun rather than the Earth with a telescope. His observations supported heliocentrism, the idea that Earth revolves around the sun, contradictory to the doctrine of the Christian Church (that Earth is the center of the universe).

He was found guilty of heresy and sentenced to house arrest. By condemning and punishing Galileo, the Christian Church silenced a potential challenge to its dominance.

The Witch Hunt during the Inquisition and the Salem Witch Trials

Obviously, magic and witchcraft have never existed and there have never been witches or wizards. Nevertheless, countless people throughout the ages espoused, though not necessarily believed in, the existence of witches who use witchcraft to secretly cause harm to other people. When they condemned certain individuals as witches, they were making false accusations. Since there were no witches and the accusations were patently false, condemnation of witchcraft never had as its goal the exposure and punishment of real witches. The real purposes of these witchcraft accusations were either to get rid of enemies, for many of the accused witches were executed, or to consolidate the high social status and power of some condemners, who were often leaders of the Christian church, through either intimidation or pretense of concern for the populace.

The Nazi Holocaust

During the Nazi Holocaust, Jews were condemned by the Nazis for blood libel (the practicing of ritual murder and sacrifice to secure the blood of Christians to use in Jewish religious rituals) , the loss of World War I, tainting the pure German Aryan blood, and many other "crimes". These false accusations led to the Holocaust, with the death of 6 million Jews. In this scenario, the purpose of the accusations was to rally the rest of the German people, the vast majority, to support Hitler and the Nazi regime before and during World War II.

Homosexuals

Sexual orientation is either determined by the genes or is set very early in development, most likely before birth, and is not alterable. All real heterosexuals should know that when they begin to be attracted to members of the opposite sex at a very young age. The same is true for homosexuals. However, instead of condemning the homosexuals of something they are born with, which would seem petty and cruel, self-righteous religious demagogues accused homosexuals of being harmful to others. Of course, in biology, homosexuals only harm themselves by potentially lowering their own relative reproductive success, and actually benefit others by taking themselves out of the game of sexual reproduction. The accusations of harm against homosexuals are therefore false accusations. The question is why the accusations are made in the first place, and we will be able to answer that question after we have explained how the behavior of condemnation evolved.

The Brain is not an All-Purpose General Thinking Machine: Evolutionary and Modular Thinking about the Human Mind

After this brief introduction on the behavior of condemnation, it's time to explore the scientific foundations that will provide us the tools to explain the behavior scientifically. When we postulate reasons and purposes for the behavior of condemnation, we are not talking about an amorphous "mind" that can somehow magically know what is beneficial or harmful and direct the "body", including the muscles that control the mouth or the hands, to issue condemnations when they are deemed beneficial and advantageous. Such an all-purpose-general-thinking-machine's existence, as has been taken for granted by social sciences in the past, is impossible because it requires an infinite amount of energy to be able to consider an infinite number of possible scenarios. We are instead talking about specific neural pathways, and networks of neurons in the brain, that are encoded in the genes and are the product of hundreds of thousands of years of human evolution. These networks and pathways, forming semi-independent but interconnected mental modules, take in specific neural inputs, integrate them, and generate specific neural outputs. The outputs in this case lead to behaviors of condemnation, and we will propose in this book how the mental module for condemnation in the human brain evolved.

Chapter 2: Biological Evolution and Evolutionary Psychology

The next two chapters will provide the scientific foundation necessary to comprehend the theory of the evolution of condemnation.

Science should be reductionistic. Every phenomenon at one level should be explained by the interactions of all the relevant elements at the level below it. Emergent properties are simply qualities of phenomenon that are surprising to us based on our inadequate knowledge of the components that supposedly create the phenomenon; it is attributed to the limits or failures of our deductive abilities, our inability to perceive the result of the interactions among all different components, and not because there is something magical and fantastical. Theoretically, it is possible to explain human social and cultural phenomena from the particles that existed right after the big bang. But we are not going so far back in time nor so small. We are only going back to the time of the existence of the first hereditary molecules. The properties exhibited by these molecules can help explain the behavior of condemnation.

All biologists agree that the theory of evolution is the foundation of biological study. However, one should go much farther. The theory of evolution is the foundation of the studies of all manifestations of life, including humanity's highest endeavors: art, music, and literature, and other human disciplines such as history, finance, government and, of course, psychology and sociology.

The basic tenets of the theory of evolution by natural selection will be briefly summarized here and then expanded to cover the ideas relevant to the discussion of the evolution of condemnation.

Biological Evolution

Before life existed, there existed organic molecules, the most important being the ribonucleic acids, or RNA. RNA molecules can make copies of themselves, but the replication process is highly inaccurate, resulting in copies with differences, called mutations. Mutations increased variation of RNA molecules in the ancient primordial sea, where life had yet to exist. Among these different RNA molecules, some had the properties of replicating faster and surviving longer than all other RNA molecules. The molecules with the enhanced abilities were able to create more copies of themselves, and since there was only a limited supply of RNA components in the environment these RNA molecules continued increasing in prevalence and eventually replaced most of the other RNA molecules. This caused a gradual change in the make-up of the different RNA molecules in the population of RNA molecules. Evolution occurred even before life existed!

Mutations in these RNA molecules continued to occur. Some mutations caused them to produce proteins, others caused them to put a double-layered lipid membrane around them and form "proto-cells", and still others caused them to yield their role as templates for replication to a different, more advanced hereditary molecule—DNA. After DNA molecules became dominant, they continued to incur mutations, produce more proteins, and eventually acquire the ability to help the proto-cell divide in two after DNA replication.

After the first organism—a simple cell—was created, life continued to evolve. One-celled organisms evolved into multicellular organisms, which evolved into the first animals, and then the fishes, amphibians, reptiles, mammals, primates, apes, and finally, humans.

Evolution takes place among living organisms in a similar fashion as among the RNA molecules in that ancient sea. First: the resources needed for survival are limited, so individuals in the population have to compete with each other, with the ones surviving able to pass on their genes. Because mutations continue to occur and, much more importantly, sexual reproduction mixes the genes of the parents to produce new combinations of genes in the offspring, new variation continue to arise. Some variants survive and reproduce better than other variants, causing their genes to become better represented in the next generation. After many generations, those traits that allowed their possessors to survive and reproduce will spread throughout the population and species so that, eventually, every individual of the same species possesses the same advantageous traits. These advantageous traits that allow organisms to survive and reproduce well in their environments are called biological adaptations, and they are the product of evolution by natural selection and are endowed in the genes.

Evolutionary Psychology

There is no doubt that some animal behaviors are biological adaptations, encoded in the genes and evolved through natural selection. In animals other than the most intelligent species (such as the higher primates and dolphins) almost all behaviors are biological adaptations. Examples of these behavioral adaptations include mating, foraging, aggression, communication, cooperation, and so on. Extending the same logic to humans, it is reasonable to assume that many human behaviors are also biological adaptations, encoded in the genes and evolved through natural selection.

The problem is that modern *Homo sapiens* evolved during the Pleistocene epoch, 100,000 years ago, as small bands of hunters and gatherers living on the grassy plains of the African savanna. Since then, the physical and social environments for humans have become vastly different and more complex, and the breadth and complexity of behaviors that all humans on Earth exhibit exploded. For example, humans now farm the land, go to school as children, read and write, play sports, listen to music, watch movies and play video games, all activities which no human in the Pleistocene, the epoch in which we evolved into modern humans, could have experienced. Our genes could not have changed much since the time of our Pleistocene ancestors, and these are still almost exactly the same genes found in bands of hunters and gatherers living on the African savanna as recently as the late 20th century. That means many of the myriad behaviors that humans engaged in throughout recorded history and all over the world cannot be simple biological adaptations per se.

If some human behaviors must be biological adaptations and yet many others, such as playing video games, cannot possibly be biological adaptations, then adaptations should not be characterized at the level of behaviors. Adaptations should be characterized at a level at which there is a one-to-one correspondence between the existence of a trait and the presence of a gene or a group of genes that code for the trait. The appropriate level to characterize adaptations must be somewhere between the genes and the human behaviors, but much closer to the behaviors than to the genes.

Once we descend below the level of behaviors, we reach the level of psychological mechanisms that produce the behaviors in response to specific external or internal stimuli.

Psychological mechanisms are networks of neurons in the brain that are organized by natural selection to perform certain tasks that are important to the animal's survival and reproductive success. All humans, with rare exceptions, share the same brain architecture, down to the most intricate wirings of the neurons, just as all humans share the same genes. Indeed, human psychological mechanisms must be the right level at which to characterize biological adaptations for human behaviors.

Psychological mechanisms respond to stimuli from the external environment or, perhaps much more commonly, the internal environment (such as other psychological mechanisms) and send signals to muscles to produce actions that we recognize as behavior. As neural networks, these mechanisms are physical structures in the brain, and they must be programed by the genes just as all other physical structures in our bodies are programmed by the genes. For example, there is a neural network in the hypothalamus that can sense the decrease in body temperature below the optimal point, and when such a condition is detected, it sends a command to various muscles in the body to begin contracting rapidly and asynchronously to generate heat in an attempt to raise the body temperature. This is the human behavior known as shivering in response to cold. There is a dedicated neural network for shivering that must have been the product of evolution by natural selection and is encoded in our genes. Similarly, there must be a vast array of neural networks, each one dedicated to its own behavioral task, that are all the products of evolution by natural selection and all encoded in our genes. These are the biological adaptations that produce human behavior. Within a human brain, they are all united by interconnecting neurons. Although all humans possess these psychological mechanisms, each one may be calibrated differently among different people, and their interactions with the other mechanisms within each brain is what produces the wide variety of behavior observed in humans.

There are several other pieces of the puzzle that we need to explain before we can attempt to propose a complete and satisfactory theory of the evolution of condemnation. These are other human behaviors that had presumably evolved prior to the evolution of the behavior of condemnation itself and whose evolution created the social environment for condemnation to evolve as a biological adaptation.

Remember, when we speak of behaviors, we are using them as shorthand for the psychological mechanisms that produce those behaviors. It's the psychological mechanisms that are the evolved biological adaptations.

Chapter 3: Communication, Deception, Cooperation, and Punishment

Communication

Communication is the sending of a signal from the signal-sender to a signal-receiver. Both the signal-sender and the signal-receiver usually belong to the same species. The signal can be in all the different modalities: visual, aural, olfactory, and tactile. When we speak of a signal, we are also including the mechanisms of signal-production in the signal-sender and the mechanisms of signal-reception in the signal-receiver, both of which are evolved biological adaptations. A signal evolved because it is beneficial to both the signal-sender and the signal-receiver. It is beneficial to the signal-sender because it can manipulate the signal-receiver to the signal-sender's advantage. If the signal does not benefit the signal-sender, the signal-sender would evolve to not send the signal. It is beneficial to the signal-receiver because the signal contains information that it can utilize to its advantage. If the signal does not benefit the signal-receiver, then the signal receiver would evolve to ignore the signal.

Examples of communication in the animal world include mating displays, calls, and sex pheromones, which use visual, auditory, and olfactory signals respectively, all of which are signals produced by members of one sex to attract members of the opposite sex for the purpose of mating. Another example is communication regarding the location of food, in which the signal-receivers obviously benefit from consuming calories and nutrients, but the signal-sender, such as a parent or a nest-mate, also benefits from increasing the chance of survival of close relatives. And still another example is the warning call about predators, which can also potentially increase the chance of survival of close relatives.

Finally, humans engage in endless communications with each other, not only through spoken and written languages but also through facial expressions, hand gestures, body postures, nonverbal sounds, bodily fluids such as tears and spit, scents, physical features such as muscles and hair distributions, and external signs and symbols, and undoubtedly in many other ways. All of these human communications have as their goal the benefit of the signal-senders, and almost always the benefit of the signal-receivers as well, or otherwise the signal-receivers would not bother to heed the signals and the signal-senders would not waste time to send the signals. However, there is an exception.

Deception

Communication is designed to be beneficial to both the signal-sender and the signal-receiver, so it is maintained in a species. Signals are not truths even though they may contain truths; they are produced and transmitted solely for the purpose of manipulation, in the biological sense. However, there is a drive for members of a species to evolve the tendency to send a signal exclusively for the benefit of the signal-sender, with not benefit accrued to the signal-receiver, or even causing harm to the signal-receiver. In other words, the signal-receiver is manipulated by the signal-sender to work for the benefit of the signal-sender at its own expense. This is deception. The signal-sender deceives the signal-receiver by sending a false signal that elicits a behavior in the signal-receiver that is beneficial to the signal-sender. The signal-receiver can always evolve the ability to detect and ignore such deceptive signals since it is being harmed by them. However, there is a gray area in which the harm is not so great that it can be tolerated, or the potential benefit of the signals is so great that they must be heeded, despite some signals being deceptive.

Examples of deceptive signals abound in the animal world, but they mainly occur between different animal species, because if deception occurs within the same species, then all members of the species can potentially be harmed because they are both signal-senders and signal-receivers. Nevertheless, one can still find plenty of examples of deception occurring within the same animal species.

Cooperation

Cooperation is the working together of a group of individuals of the same species to achieve a goal that none of the members can achieve alone, or with the same efficiency or productivity. Cooperation can only occur among social animals that recognize each other and are intelligent enough to work together, each contributing effort to achieve the common goal of the group. All the members of the group are willing to pitch in because they will all derive a greater benefit than if they were to work alone.

Cooperation in hunting is observed in social predators such as wolves, hyenas, wild dogs, orcas, and chimpanzees. Cooperation evolved among the ancestors of humans, most likely the common ancestors of humans and chimpanzees, but it evolved far beyond just cooperation in hunting. Even among the chimpanzees, making the fair assumption that they resembled our common ancestors more so than humans, cooperation among males occurs in the obtainment of females, territory, food, and social status. Among humans, cooperation can occur in every imaginable endeavor in which the combined efforts of multiple individuals would achieve better results than the effort of one individual alone. Cooperation is observed in team sports, politics, science, and warfare. It is ubiquitous in human society.

However, after cooperation has evolved, cheating can easily evolve as well. Cheaters do not contribute the requisite efforts to the cooperative endeavor, and yet they derive benefit from all the efforts contributed by other members of the group. Without contributing efforts, the cheaters can save energy and time for something else that increases their fitness. We have all encountered such cheaters in our lives. They are the idlers, lazy bums, and loafers who did nothing while others worked and yet reaped the resulting benefit once the tasks were accomplished. Such cheaters can cause any cooperation to collapse. If one can do nothing and get the same benefit, why would anyone do anything? Such cheaters are harming those who do the honest work. Fortunately, there is a way to make sure that partners in cooperation do not have an extra incentive to cheat.

Punishment

Punishment is aggression toward an individual or group that poses a threat or has harmed others. Harm can take many different forms, but the end result is always the same—a potential decrease in survival and reproductive success in the victims. For example, cheaters in a group pose harm to others who contribute real effort, time, and energy in a cooperative endeavor. Others can render harm more directly, such as through theft, robbery, battery, rape, or murder. Punishment is a behavior that evolved as a way to deter future harm from not only the malefactors who have caused harm or who posed a threat, but also from the bystanders, so they would refrain from causing harm in the future.

Punishment is harm to the malefactors, a loss in resources—such as wealth, health, lifespan, or even life—so severe that the malefactors and others would not risk it by causing harm in the future. Like harm in general, punishment can take many forms: payment, loss of property, public humiliation resulting in a decrease in social status, confinement away from normal and desired social interactions, ostracism, physical discomfort, psychological discomfort, bodily injury, and death. Death as the ultimate form of punishment obviously serves not to deter the malefactor himself, but to deter others who have observed or are aware of the capital punishment. Death, of course, also eliminates the malefactor entirely, so he would have no opportunity to cause harm in the future. It is reserved for the most egregious of harms.

Harms have been committed throughout history and in all cultures. The tendency to cause harm to others is engrained in the human psyche and encoded in human genes. Likewise, punishment, in all its various forms, has been engaged in by people everywhere on Earth since time immemorial, ever since humans first began committing harm on others. One could say that punishment exists because the desire to harm others has been such a powerful human drive.

Cooperative Punishment

A punishment has a higher probability of being strong and effective if it is meted out by not just one individual, but by a group of individuals, all of them contributing or at least consenting to the cooperative punishment. As cooperation and punishment had already evolved in humans, cooperative punishment was but the natural next step in human evolution, setting up the stage for the evolution of condemnation.

Chapter 4: The Evolution of Condemnation

Condemnation is the accusation that an individual or group is harmful. What are the harms for which these accusations are issued? The harms must have decreased survival and reproductive success significantly in humans' evolutionary past and therefore constituted enormous selective pressure for the evolution of biological adaptations to counteract them. They include behaviors that are condemned in all human societies: lying, cheating, theft, robbery, adultery, rape, assault and murder. There may be many more behaviors that are considered harmful and therefore condemned, at least by some groups.

As we learned from the previous chapter, one biological adaptation against harm by other humans is punishment, which is so costly to the malefactor that it serves as a deterrent for similar harm in the future. The effectiveness of the punishment corresponds to its magnitude: the stronger the punishment, the more effective it is. In a group of humans, when an individual is harmed by a malefactor, he can exact punishment on the malefactor himself, thus deterring the malefactor from committing the same or similar harm against him in the future. However, as an individual, his power to exact punishment is necessarily limited. He is one person going against another person, the malefactor. Unless he is much bigger, stronger, or more skilled at wielding a weapon, his punishment may not cause too much harm on the malefactor and may not amount to much of a deterrence. However, he is an individual living in a group, and humans, as social animals, have already evolved the ability to cooperate. If he could tap into the combined strength of the group in punishing the malefactor, the punishment should be much stronger and much more effective as a deterrent. Therefore, there must have been a tremendous selective pressure for the evolution of a psychological mechanism in humans that compels victims or potential victims of harm to solicit the cooperation of other individuals in a social group to punish the malefactor, because the successful avoidance of future harm from malefactors must be highly beneficial to the survival and reproductive success of the victims or potential victims.

As social animals, humans had already evolved cooperation in combating enemies and rivals early in their evolution. They had already evolved psychological mechanisms to detect harmful individuals and to trigger hostility and aggression toward them. As a victim or potential victim of harm from a malefactor, a human can tap into the combined strength of the members of a social group by activating their psychological mechanisms that detect malefactors, triggering their hostility and aggression toward the malefactor. In other words, he could elicit the cooperative punishment of the malefactor by issuing a condemnation.

Therefore, the function of condemnation must be to elicit cooperative punishment from other group members. When a person is harmed, or in danger of being harmed, by a malefactor, besides exhibiting his own hostility and aggression toward the malefactor, which serves as punishment and aims to deter future harm, he can also issue a condemnation to other members of his social group and solicit their cooperation in punishing the malefactor. Again, punishment carried out by a group, even if it is only a group of two, should be more effective than punishment conferred by a single individual. Since harm caused by malefactors has been ever present in human groups, there must have been a great selective pressure for the evolution of a drive to alert other humans of the presence of a malefactor in a social group, once the malefactor is detected, and especially when the malefactor has harmed or possesses the potential to harm one.

In short, the accusation of harm or potential harm in a malefactor issued to an audience is the behavior of condemnation, and the words or phrases that convey this is the signal of condemnation.

In primitive hunter-gatherer societies, the signal of condemnation could be issued for any harm or potential harm that hunters might experience in their lives. For example, if a hunter detected another community member stealing his food, he could shout, "Stop, thief!" If he witnessed the murder of another, he could shout, "Murderer!" As human societies progressed in complexity, so did the condemnation that humans issue grow in sophistication. Indeed, condemnation may often be delayed so that the condemner can elicit stronger and more effective cooperative punishment. In our society today, when a person detects wrongdoing in another, he may not issue a condemnation right away but hold the information for a long time, until he has gathered sufficient evidence to convince his audience when he issues his condemnation, and therefore his condemnation is more likely to obtain its intended effect.

Condemnation, as the issuance of a signal to alert the receivers of a common malefactor that poses harm or threat to both the condemner and the audience alike, first evolved as an honest signal. Both the condemner and the audience can potentially be harmed by the malefactor, and therefore condemnation, if it elicits cooperative punishment from the audience, benefits both the condemner, the signal-sender, and the audience, the signal-receivers.

However, as established in the previous chapter, after an honest signal has evolved, it can be taken advantage of by being used for deception, so that the signal-sender can reap the benefit of the signal without paying the requisite cost or sharing the benefit with the signal-receivers.

How can the honest signal of condemnation be misappropriated to use as a deceptive signal to benefit the signal-sender but not the signal-receivers?

To answer this question, let's examine the possible benefits of a signal of condemnation to a condemner. First of all, condemnation elicits cooperative punishment of the condemned, a malefactor who has harmed or is likely to harm the condemner and the audience, and the punishment often decreases or obviates future harm from the malefactor. This benefit is shared by both the signal-sender and the signal-receivers, and therefore the signal is an honest signal. It is because of this shared benefit that both condemnation and receptivity to condemnation have evolved.

There are potentially two other benefits to the condemner from a signal of condemnation. First, the malefactor who posed threats or committed harm to the condemner is likely a social rival of the condemner. All human behaviors, including the behavior of condemnation, evolved in small, tight-knit social groups where everyone knew each other and any two individuals were likely to be social competitors whose interests were often in conflict. If a condemner's condemnation led to the cooperative punishment, whether justified or not, of a malefactor who was also a social competitor of the condemner, then it could have reliably led to a reduction of the social rival's survival and reproductive success, and as a consequence an increase in the condemner's own survival and reproductive success. This creates an enormous selective pressure to condemn a social rival for the purpose of eliminating competition from the social group, rather than to avert future harm to both the condemner and the audience. Eventually, this drive to use condemnation against social competitors could have been so successful that it spread through the human populations and became a species-wide behavioral and biological adaptation.

However, such a form of condemnation is essentially a deception. The condemner issues a signal to inform others of a malefactor's harm or threat to them all, but he actually tries to elicit cooperative punishment against a social competitor. Oftentimes, such a condemnation is a de facto deception: the condemned social rival is actually not harmful or threatening, and the accusation against him has been fabricated by the condemner. The question is, why did the human species allow such a deceptive behavior to evolve, if deception could potentially harm everyone? Why did humans not evolve the ability to see through the deception and refuse to respond to such a condemnation? There are several possible reasons. First, much of the signals of condemnation may still be honest signals. There may still be a greater proportion of honest signals compared to deceptive signals. Humans simply cannot afford to stop responding to these signals, despite some of them being deceptive. Second, not responding to a signal of condemnation may have dire consequences that humans simply cannot afford. So, it is better to still respond to signals of condemnation even if most of them turn out to be false and deceptive. Third, the condemners have evolved the ability to make their condemnations appear genuine, either by discovering actual harm or threat on the part of their social rivals, or by carefully building up a case against their social rivals before issuing the condemnation.

The second additional benefit to the condemner from a signal of condemnation is that it can potentially elicit social support from the audience and thereby elevate the condemner's own social status. When the condemner issues an accusation, not only is he advertising the harmfulness of the condemned, but he is also advertising his hostility toward the condemned and his willingness to join a coalition to punish the condemned. Moreover, as someone who appears to care about the welfare of the potential victims of the condemned, he advertises his goodness and value to the group. All of these effects, taken together, would likely cause his audience to harbor good-will toward the condemner and offer social support, leading to an elevation in the condemner's social status (a purpose for which condemnation may also have evolved).

To summarize, the following are biological functions of the evolved behavior of condemnation. All of them confer benefits on the condemner, the signal-sender. Only the first one is an honest signal that also confers benefit on the audience, the signal-receivers. The other two are deceptive signals that only confer benefits on the signal-sender and not the signal-receivers.

1. Condemnation elicits cooperative punishment from the audience against a malefactor who is harmful or potentially harmful to both the condemner and the audience, leading to a decrease or termination of the harm conferred, or threat posed, by the malefactor.
2. Condemnation elicits cooperative punishment from the audience against a social rival, decreasing the social rival's status and otherwise harming the social rival so that the condemner gains a social competitive advantage.
3. Condemnation elicits social support for the condemner, increasing the condemner's social status.

The strongest evidence for the existence of an evolved psychological mechanism is the emotions, especially the strong ones, associated with its activation. Emotions are the concerted activation of neural networks in the brain, encoded in the genes and produced by natural selection. If condemnation and the reception of condemnation are both reliably associated with the emotions of righteousness, indignation, anger, fear, hostility, and hatred, there must have evolved psychological mechanisms in the human brain dedicated to both behaviors. Therefore, condemnation must be an evolved biological adaptation.

Chapter 5: Culture

Condemnation makes up a sizable portion of the human cultural fabric. It is embedded in human morality, religion, legal systems, politics, news, and in many other cultural mediums. Before we talk about condemnation in culture, let's define and explain the phenomenon of culture itself.

Culture is a difficult concept to delineate and grasp. Culture may be loosely defined as the creating, spreading, and acquiring of novel concepts among social animals. In that sense, the word "culture" refers to the process of concept transmission rather than the content of what is being transmitted—the cultural artifacts or products. Of course, the term "culture" also refers to the collection of all artificial elements produced by humans, including ideas, theories, books, songs, movies. It is something that is only possible in brains intelligent and powerful enough to understand and remember ideas, and it is facilitated by verbal language and writing. Indeed, the impetus for the development of culture must have been the existence of complex brains with good memory.

Why would humans create culture in the first place? Culture is simply communication across long periods of time and over large distances through mediums such as human brains and physical objects, the latter of which include painting and writing on walls, stone tablets, and paper. Before there were means to physically record culture, such as writing, drawing, and the printing press, culture was at first transmitted from one human brain to another through verbal communication. Afterwards, cultural has been recorded in much more permanent forms instead of the human brains which are liable to distortions and forgetfulness. However, despite the physical medium in which culture can now be recorded, the mechanism of cultural transmission is still the same as when only human brains were available. Culture is recorded in either the neuronal connections of the brains or a physical medium so it can attain a greater longevity and be transmitted over greater time and space. Other humans can access this recorded culture, so their brains can be influenced in ways that are intended, and they can decide to pass it on verbally or through recording in physical medium again.

In other words, culture is merely a more sturdy and far-reaching form of communication. Culture affects the brain like direct signals such as human verbal communication. As communication, its function is the same as any other type of communication in the living world: to manipulate the signal-receiver to the benefit of the signal-sender. However, besides the signal-sender and the signal-receiver, culture gives rise to a new player in the game of communication: the signal-transmitter. The signal-transmitter is not a static agent that simply transmits whatever it receives to others; it is an active participant who can modify the signal that it is transmitting to its own advantage. The signal that is transmitted does not remain the same as it was when it was produced by the original signal-sender. Instead, every time the signal is transmitted by a signal-transmitter—and there could be a vast number of signal-transmitters—it can potentially be altered, either subtly or drastically, to the benefit of the signal-transmitters.

The cultural elements that exist in the brains of humans or in physical copies in books, recordings, videos, or computer files, may have been transmitted and re-transmitted countless times. They could be beneficial to some and detrimental to others. Their predominance or lack thereof depends on the number, power and avidity of the humans who transmit them. Indeed, these signal-transmitters, temporary vessels of cultural elements, can actively manipulate culture to their advantage. Humans possess evolved psychological mechanisms that evaluate each cultural element stored in their brains to measure its impact on their own fitness. If they judge that the cultural elements are beneficial to their survival and reproductive success, then they support them and try to propagate them. If they judge that the cultural elements are detrimental to their survival and reproductive success, then they renounce them and try their best to suppress their spread. Highly prevalent cultural elements have been judged by great numbers of humans, or in some cases a few powerful humans, to be beneficial to them.

Culture is actively maintained by humans to their own advantage.

Some of these cultural elements are seemingly transmitted with great ease because the human brains eagerly embrace them and just as eagerly pass them on. These are labelled memes, a term coined by the great evolutionary biologist Richard Dawkins in his seminal work *The Selfish Gene.* Originally, the term "meme" refers to any culture element that can potentially be passed on from brain to brain. But lately, it has come to mean a photo or a video that is spread rapidly through the internet and social media, having gone viral, so to speak. Indeed, the term "meme" has become a meme.

Below we will present an example of cultural elements that are propagated for the advantage of those who engaged in their propagation.

Stories

Humans live in social groups in which other members, which may be friends, enemies, lovers or rivals, may have large impacts on one's survival and reproductive success with their actions and behaviors, and therefore one needs to know as much about others in the social group as possible in order to stay ahead of the game and outcompete others. For this reason, humans have evolved a penchant to hear about other people (what they do, who they like and hate, who they are dating, and so forth). They welcome stories about other people with boundless curiosity and eagerness, making journalism the second oldest profitable profession after prostitution. Some humans with great imagination decided to satisfy this desire for true stories with fake stories, which have no less power to satisfy the hunger for stories, and indeed even more so because these stories could be wildly embellished with exotic and magical details far beyond an ordinary human's normal experience. And thus the storyteller is born, likely the third oldest profitable profession.

The storyteller spun tall tales of the wildest imagination throughout the ages. Some were truly gifted, and their stories were passed down through the generations, first orally and then recorded on tablets, scrolls, and books. Later incarnations of the storytellers include the imaginative historians, the poets, the playwrights, the novelists, and the moviemakers. Humans throughout the world enjoyed reading, hearing, and watching these stories with the same fervor that they had listening to the original gossipers in the darkness of caves. Taking advantage of people's predilection to hear stories about real people, the storytellers weave fictional tales that held people spell-bound. And these storytellers gained social prestige for their vivid imagination and storytelling abilities, just as the original gossipers must have earned tremendous respect for the wealth of knowledge about other humans that they deigned to share.

Stories are simply elaborate signals. Just like all other signals, stories have the potential to manipulate others to the storytellers' advantage. One such advantage is the prestige that humans confer on storytellers with great imagination. But there is another potential benefit. These stories can be used to propagate a belief in the superiority of some individuals or groups over others. Because humans have a tendency to believe in stories, even stories that have no chance of being true, such stories would confer on the celebrated individuals or groups higher social status and therefore greater survival and reproductive success. Obviously, the storytellers themselves belong to the superior groups that they are touting. Examples of stories that propagate a belief in the superiority of certain groups include the myth of God's chosen people in Judaism, and the superhero movies of early 21st century in which almost all of the superheroes were Europeans or descendants of Europeans, even those supposedly from outer space.

Chapter 6: Morality

Humans throughout the ages have grappled with the concept of morality. Philosophers such as Confucius, Lao Tzu, Mung Ze, Socrates, Plato, Aristotle, and Nietzsche expounded on it; religions such as Buddhism, Judaism, Christianity, and Islam have preached on it; novelists such as Tolstoy and Dostoevsky explored it.

Many people believe that there is a morality "out there", pure, perfect, immutable and unassailable, that every human should follow. Others, especially the academics, believe morality is culturally relative, and one should not hold one's morality over others, and we should just agree to disagree on morality.

Indeed, morality is one of the most profound subjects of all. If science purports to be able to explain anything in the world, including human cultural phenomena, then the scientific theory that can get to the bottom of morality should be considered one of the greatest achievements in science.

To understand morality from the point of view of science, one needs to really understand what it is. The best way to do so is to examine all the manifestations of what we perceive as morality. Morality consists of two components. The first component is the set of human behaviors that are considered bad and evil and deserving of punishment. The second component is the set of human behaviors that are considered good and worthy of praise and emulation. The second component actually occupies only a small fraction of the totality of human morality, even though when we talk about being "moral" we often refer to the second component of morality and equate being "moral" with being good. Nevertheless, those who do not conform to our culturally accepted conception of good are therefore consigned to the category of bad or evil, to however slight or large a degree. In other words, those who fail to follow the code of good behavior are regarded as bad and deserving of punishment just as those who committed crimes or sins.

The first component of human morality, the human behaviors that are considered bad or evil, includes the behaviors that have almost always been designated as such in all human cultures through time immemorial: lying, cheating, stealing, robbery, infidelity, rape, violence, and murder. However, a caveat is that the designation of such behaviors as bad or evil may only apply to humans regarded as belonging to the same social group, which can be a tribe, village, city, nation, ethnic group or race. With respect to someone from a different social group, such deeds done to him or her may no longer be regarded as bad or evil, and therefore is not deserving of punishment and elicits no guilt or remorse. From the point of view of the government of a multi-ethnic nation, this is why it is so important to downplay ethnic and cultural differences and to emphasize the sameness of all humans to prevent one ethnic group from committing such "crimes" against another.

The first component of human morality also includes some behaviors that are more fluid in terms of their designations, such as homosexuality and infidelity. Homosexuality had been considered a sin and had once been a crime punishable by imprisonment or death (and may still be so in some parts of the world) but in the early twenty-first century homosexuality is widely accepted throughout the civilized world. As for infidelity, there were two American presidents who underwent impeachment trials two decades apart; one was tried for lying about infidelity, and the other had countless infidelities and lied about them, but was tried for something else. Early in the 21st century, infidelity and adultery are not crimes in the majority of U.S. states, a far cry from the times when adultery might have been punished by death.

Morality is a cultural complex, with various cultural elements created and, much more frequently, transmitted by people's brains to manipulate others for the benefit of the cultural creators and transmitters. Because of the fact that the designations of certain behaviors are flexible, sometimes regarded as immoral and sometimes not, there has been a long-held belief among academics that morality is relative. Each group of people may have its own morality, and we just have to accept that moralities of different groups may differ, and it's a worthy endeavor just to document all the differences. These academics approach the study of human morality as though the human mind is an all-purpose general intelligence capable of producing all manners of behaviors imaginable, without any scientific proof. Again, such an all-purpose general intelligence requires an infinite amount of energy to consider an infinite number of different scenarios and cannot exist in the universe, let alone inside a human brain.

Simply postulating that all cultures are relative is not the way scientists should approach the study of morality. Despite the fact that the designations of behaviors as immoral may be flexible from period to period and from culture to culture, there is something that is not relative: human nature. The human psychological mechanisms that are the products of evolution by natural selection are all the same among all members of the human species throughout history and in all cultures. Therefore, as scientists, we must believe that there are scientific explanations for the variation of moralities among different groups of people, and the scientific explanations must lie in the biological adaptations that generate the human behaviors in the first place, namely, the psychological mechanisms.

As we have established, the one common theme of all moralities, without exception, is the stipulation that certain behaviors should be condemned and punished. Even moralities that extol certain behaviors as good and commendable are casting the opposite behaviors as bad and condemnable. And as we also have established, condemnation is an evolved human behavior controlled by an evolved psychological mechanism. Let's take the natural next step in our postulation on the origin of human morality, taking into account the intimate connection between the cultural phenomenon of morality and the psychological mechanisms behind condemnation, and offer a psychological and cognitive explanation that connects morality to condemnation: Human morality is a collection of cultural elements created and transmitted for the purpose of allowing the behavior of condemnation to be employed to the advantage of the creators and transmitters of those cultural elements.

Instead of postulating that the human brain is an all-purpose general thinking machine, this hypothesis begins with two well-established biological adaptations in the human brain: the psychological mechanisms for condemnation and for cultural propagation. It then posits the existence of neural connections that can evaluate the cultural elements to determine whether they would allow the activation of the psychological mechanism for condemnation, a behavior that is beneficial to the condemner. If so, the brain propagates these cultural elements. If not, the brain does not propagate those cultural elements, and even suppresses them.

Stripped of all the pretensions and obfuscations, morality is simply a set of cultural elements that stipulate which behaviors are worthy of condemnation. It is constantly modified for the benefit of the most numerous, or the most powerful, or the most crafty, or the most ardent, depending on which one ends up winning the day. There is no supernatural authority dictating the right versus wrong, and there is no ultimate morality that can be reasoned out and that can please everyone. At the end of the day, many people's interests are in conflict with each other's, and what is beneficial for one is often harmful for another.

However, morality is often couched in languages of absolute rightness and absolute authority. This is intentional and is designed to minimize objections, protests and conflicts. What could be better to reinforce one's brand of morality than to invoke a supernatural supreme being as its originator who demands absolute obedience? Never mind that such a supernatural entity does not exist. As long as enough people believe in its existence, it will do the trick, because the sheer number of believers can silence the others into submission in deference to the large presence of believers.

Let's examine morality with respect to the phenomenon of homosexuality. Homosexuality has the unique characteristic of being not harmful to anyone except the homosexuals themselves, and yet has often been condemned throughout history and in most cultures. Homosexuals had been condemned to imprisonment and death, they are condemned by many religions, and there had been laws against homosexuality even in the most progressive countries until relatively recently. How do we explain a morality that condemns homosexuality as harmful when, in fact, it is not?

We have to invoke the second and third functions of condemnation, condemnation as a weapon against social rivals and condemnation as a means to raise social status. The human mind so readily accepts homosexuality as maleficent and worthy of condemnation because it allows many people to condemn this perceived small group of individuals, the homosexuals. Homosexuals were condemned not because they were harmful, although the condemners claimed they were. They were condemned because doing so caused the homosexuals to have lower social status and the condemners to have higher social status. And there was no fear of retaliation from the homosexuals because they were perceived to be a small minority possessing little to no power.

However, the situation quickly changed with widespread electronic communications. The visibility of homosexuals grew. In many people's minds, the group went from a small minority that can be condemned with impunity to a sizable group that frequently impinges on people's awareness. That was when homosexuality changed from being a condemned group to one welcomed ostensibly in the name of diversity.

In summary, morality is a set of cultural elements constantly reinforced by cultural transmission to manipulate other people's minds to the moralists' advantage. There is no objectivity to morality, only subjective self-interests, even though people try to create the illusion of objectivity so they can derive authority from this morality when they use it to condemn. Besides claiming that there is an objective morality, many people created a supernatural deity to lend authority to their brand of morality, so that they could use it to condemn without repercussions. But ultimately, whatever morality one believes in is purely for selfish reasons: maybe the people who uphold a brand of morality are too strong and powerful and therefore one cannot help but subscribe to it; maybe the morality allows one to condemn the people who are really harmful, and maybe it allows one to lower other individuals' or group's social status and elevate one's own social status.

During human social development, one of the tasks of the brain is to learn which cultural elements in the social environment fit the mold of being condemnation-worthy and remember it. The cultural elements may be endlessly fluid and variable, but the mold that they fit in is not, because it is a biological adaptation evolved by natural selection and encoded in the genes. The exact shape of the mold may depend on the rest of the brain and the identity of the individual. For example, the cultural element that homosexuality is evil likely will not fit the mold for condemnation for a homosexual, no matter how strong a signal it is. The person will not allow himself to be condemned. This is why whenever someone argues that homosexuality is not morally wrong, that person immediately invites suspicion of being a homosexual, even though he or she may merely be an objective scientist.

The brand of morality that exists in a society is not random or stochastic but depends precisely on who the most numerous or the most powerful are in that society. For example, unabashed greed may be deemed evil in certain societies, but praiseworthy in others. Adultery may have led to presidential impeachment at one time in American history but barely stirred a ripple in the political world at another moment in time. When we study morality, we should not simply say that morality varies from culture to culture and leave it at that. We should delineate exactly the psychological mechanisms involved in maintaining the collection of cultural elements, namely moral dictums and precepts, that constitute morality in that particular culture or society.

Chapter 7: Religion

Religion is culture taken to the extremes. It is a cultural phenomenon with myriad inconsistencies. It is a cultural system constantly maintained by different groups of people, each with its own agenda and benefits to be gained from the specific religion that its members support. Much of religion is scientifically nonsensical. Paradoxically, the nonsensical elements, such as the father and the son being one and the same and a man rising from the dead, make them more lasting in people's minds because they are unexpected and striking and therefore easy to remember. The nonsensical elements and the elements that benefit various groups maintain an uneasy cohesion throughout the ages, taking residence in people's minds as well as in books and other media.

The question is: Which groups of people benefit from religion?

There are actually many different groups of people who can benefit from religion, groups that may not have much connection with each other; however, their combined support of a religion keeps religion going strong.

Among these disparate groups of people is one group whose members are drawn to the moral aspect of religion, and they seize on opportunities afforded by their religion to engage in the behavior of condemnation. They have their own brand of morality, but by attaching it to religion, with the authority of the religion behind them, they encounter less resistance. They could always say that their own brand of morality did not originate with them but with the supernatural deity that many other people also believe in for their own reasons. If anyone argues with their brand of morality, he or she would be branded a heretic and sinner and can be condemned and punished, until no one dares to raise a voice against their brand of morality so they can then condemn at will.

Let's take the example of Christianity.

Christianity is a complex religion with many different components that may be supported by different swaths of humanity, who supported Christianity for their own unique reasons, but they collectively help to sustain Christianity as a cultural phenomenon. For example, some people support Christianity because it reaffirms the superiority of a certain race, namely the descendants of people who lived in Europe. Others support Christianity for a sense of belonging because of Christianity's large presence throughout the world and its tradition of reaching out to people who are alone and lonely and who need companionship. In the past, Christianity appealed to the poor because it claimed that the poor shall inherit the earth and it also had a tradition of helping them, providing food and shelter for the homeless. Christianity also appeals to the group mentality of some people, and joining Christianity make them feel the strength in numbers. And many people find Christianity appealing because it gives them a sense of moral superiority, as though just by claiming to be Christians they already are higher in social hierarchy compared to the non-Christians.

We are not merely saying that there are simply multiple reasons for joining Christianity and leave it at that. We are saying that there are multiple evolved psychological mechanisms that are involved in the tendency of some people to join Christianity.

For many, they support Christianity to be able to condemn. One of the benefits of Christianity is that it contains many dictums that allow people to condemn many other groups of people. Indeed, Christianity can be characterized as the religion of condemnation. Since we have established that condemnation is beneficial, Christianity confers potential benefits to many condemners and is maintained as a powerful and influential cultural system by them. For example, throughout history, Christianity allowed Christians to condemn greed, sex, infidelity, prostitution, worship of other gods, homosexuality, transgenderism, and so on and so forth. Christianity even condemned people for not believing in Christianity, as in Mark 16:16: Whoever does not believe will be condemned. Of course, the condemners have not committed any of these behaviors or belonged to any of the groups that were condemned, so they would not be harmed by their own condemnations, very conveniently.

Let's take the example of abortion. The aborted embryo or fetus was the potential offspring of the mother who decided to get the abortion. If there was any harm, the harm occurred to the mother because of the lack of gain in her reproductive success, although she probably already had or will have many more spontaneous miscarriages, so this harm was very minor. The human embryo or fetus never appeared in our evolutionary past as something that could have influenced our evolved human psychology. It was never supposed to be regarded by humans as a human life, the same way that you, the reader, or me, the author, is regarded by the evolved human mind as another human life. The killing of you and me by a murderer would constitute great harm and should be soundly condemned, or at least I hope so. However, the Christian anti-abortionist moves to foist the concept of "human life" onto a fetus, or an embryo, or even a zygote, a single cell, as they oppose scientific research using discarded embryos. This is obvious deception. While they are arguing against abortion, they could have spent their time fighting for the rights of real human lives among the impoverished, the refugees, the gangs. They do not condemn males, including many anti-abortionists, who ejaculated millions of sperm everyday even though each sperm is basically half of a zygote and has the potential to become a real human, if it fertilizes an egg. They do not condemn people, including all anti-abortionists, who lose millions of cells everyday through their skin and digestive tracts, even though with potential genetic technology each of those cells can be turned into a cell that is just like an embryonic stem cell, and eventually, as genetic technology continues to advance, each of those cells can potentially be used to produce a human clone, just like the sheep clone Dolly. No, they would not condemn themselves. They would only condemn girls or women with unwanted pregnancies that could have resulted from rape, ignorance, or impulsivity, and they would condemn doctors who are preventing these girls and women from having children they did not want and who are preventing people from being born to mothers who do not want them.

In short, the most zealous and self-righteous condemnations are most likely false and deceptive.

Epilogue

Great scientific ideas attempt to answer the most profound and mysterious questions that have puzzled humankind for millennia: Why does the universe exist? Why are we the way we are? What is right or wrong? When we assess which scientific ideas are among the greatest of all time, we consider how profoundly and powerfully they changed our way of thinking, how revolutionary and innovative they are, and how vast is the scope of their applicability and influence. Judged on those criteria, the greatest scientific ideas include classical physics by Newton, the theory of evolution by natural selection by Darwin and Wallace, the particulate nature of genetic inheritance by Mendel, special and general relativity by Einstein, quantum mechanics, and evolutionary psychology, the last two being the products of group efforts. To that list we must add the theory of the evolution of condemnation, which completely and unequivocally answers one of the most profound questions of all, "What is right or wrong?" With this theory, human morality has finally been scientifically explained, with no room for equivocation. It is the direct product of DNA replicating itself with less than perfect accuracy. The theory of the evolution of condemnation is a scientific theory that explains not only human morality, or more precisely human moralities, but also human legal systems and human religions. Because the behavior of condemnation is often deceptive and harmful, and it undergirds human moralities, legal systems, and religions, it is devoutly hoped that the light cast on the behavior of condemnation and the role it plays in various social phenomena can significantly decrease unnecessary human suffering in the future.

March 10, 2022

Made in the USA
Monee, IL
08 November 2026

67972901R00038